M & B Technical Library TL/ME/3

General Editor: J Gordon Cook, PhD, FRIC

Plasma Spraying

R F Smart, BSc, PhD, AIM
Manager, Materials Department

J A Catherall, BSc, AIM
Head of Materials Technology Group
Associated Engineering Developments Ltd

Mills & Boon Limited
London

First published in Great Britain 1972
by Mills & Boon Limited, 17–19 Foley Street
London, W1A 1DR

© R. F. Smart and J. A. Catherall 1972

ISBN 0.263.05074.2

Made and printed in Great Britain
by Butler & Tanner Limited, Frome and London

PLASMA SPRAYING

All published by Mills & Boon Limited

Frontispiece. Plasma spraying.

CONTENTS

Preface

This book briefly describes the principles and practice of plasma spraying, a rapidly developing science which retains more than a trace of art. It is intended to give a general description of the process, together with an assessment of the present state of the technology. The treatment is, of necessity, brief but a more detailed account of the various aspects of the subject may be obtained from the references listed in the bibliography. Specific references in the text have been considered unnecessary but an extensive literature survey may be found in Reference 2 in the bibliography.

The authors wish to thank the Directors of Associated Engineering Developments Ltd. for permission to use photographs and experimental data in this monograph. They are particularly grateful to their colleague Mr. F. J. Atkins for his constructive comments on the text.

1. Introduction

Surface treatments have been used since the earliest days of metalworking and in modern engineering practice their application has become widespread. They range from conversion coatings, which involve a sub-surface reaction with the base material, to deposition coatings, in which the deposit is attached to the base with little or no reaction with it. The distinction between the two lies essentially in the bond, which in the former case is chemical and in the latter is physico-mechanical in nature. Many surface treatments, of course, fall between these extremes.

Plasma spraying is basically a deposition process and is one of the group of coating techniques known originally as metal spraying but now, more correctly, as thermal spraying. Although differing in detail, all generally involve three features:

   (i) the material to be sprayed is heated so that it is substantially molten,

   (ii) the molten material is projected onto the base material (or substrate),

   (iii) the projected material adheres to the substrate to give the required coating.

The idea of spraying metals may be traced to the development of atomisation techniques for the production of metal powders and particles and to the realisation, by Dr. M. U. Schoop at the beginning of the century, that by directing the particles at a suitable target an adherent and usable coating could be produced. During the next two decades Schoop and his associates developed the method and produced a variety of equipment designs. Based largely on these developments, commercial spraying was introduced in the 'twenties; it has since achieved wide usage and now supports an extensive coating and hard facing industry. A number of variants of the technique are used (see Table 1) but, for commercial purposes, they

11

may be divided into four main types:

(a) flame spraying
(b) arc spraying
(c) detonation spraying
(d) plasma spraying.

Table 1. Thermal Spraying Techniques

A Chemical flame	i Oxy-gas: (a) wire/rod
	(b) powder
	ii Rocket expulsion: powder
	iii Detonation spraying: powder
B Electrical heating	i Resistance: atomisation
	ii Induction: wire
	iii Arc spraying: wire
	iv Plasma arc (non-transferred):
	(a) wire
	(b) powder
	v Plasma arc (transferred): powder

The distinguishing features of the four processes are shown in Fig. 1.

The bulk of commercial production is based on flame spraying, using oxygen–gas mixtures as the fuel and powder or, more usually, rod or wire as the feed. With proper precautions, a fairly dense and reasonably adherent coating can be produced. Bond strength and density can be increased by using the torch to fuse the deposit and the coating material may contain flux additions to facilitate this. The equipment for flame spraying is cheap and, consequently, capital costs are low. However, the process suffers from two major limitations: firstly, the flame temperature is insufficient to melt the more refractory metals and many ceramics; secondly, the substrate tends to be heated by the flame and this may cause difficulty with workpiece distortion, as well as precluding the use of low melting point substrates (e.g. plastics).

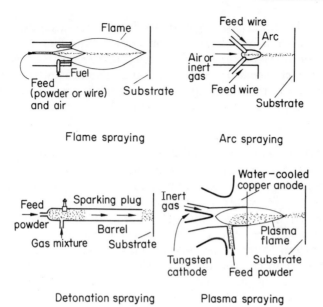

Flame spraying

Arc spraying

Detonation spraying

Plasma spraying

Fig. 1. Features of thermal spraying processes.

A complementary process to the flame method is arc spraying, in which the material to be sprayed (which must be available as rod or wire and be electrically conductive) forms the consumable electrodes of the gun. The molten particles are blown, usually by an air blast, against the substrate and, since the gun can reach a slightly higher temperature than the oxy-acetylene flame, marginally denser coatings are produced. Against these advantages, however, arc spraying has a higher capital cost although running costs are claimed to be slightly lower.

During the past 10–15 years, the needs of the aircraft, aerospace and electronics industries have stimulated the development of processes to give coatings of generally higher quality and to extend the range of spraying to the refractory metals and ceramics. These requirements have led to the commercial introduction of detonation and plasma spraying.

The detonation process (or flame plating) was developed from studies of flame propagation in tubes. The coating

B

powder and a carrier gas are fed into an oxy-acetylene gas chamber in which detonation occurs some four times a second; this produces high velocity impingement and gives very dense coatings with excellent adhesion. However, the process is the most expensive of all the thermal spraying techniques and has more geometrical limitations than the others.

Plasmas were used over 50 years ago to synthesise nitric oxide from air but it was only during the 'fifties that the technology was developed to make them commercially useful. To create plasma conditions, a suitable gas is passed through a high-current arc; at sufficiently high temperatures, the gas becomes ionised and collisions between electrons and ions generate radiant energy. In a plasma torch, the thermal balance is changed by constricting the arc, and this raises the temperature substantially, usually to 15,000–20,000 K.

Fig. 2. Plasma spraying.

A typical plasma spraying assembly is shown in Fig. 2. The heart of the system is the plasma torch (or gun) which is shown schematically in Fig. 3. The electrodes are contained within the gun and are shaped to give the required constriction while the arc gas and the feed material (generally powder) are fed in through different ports. The operation of the set and the efficiency of the gun, as well as the quality of the deposit, depend critically on such factors as electrode design, powder size, substrate preparation, etc.

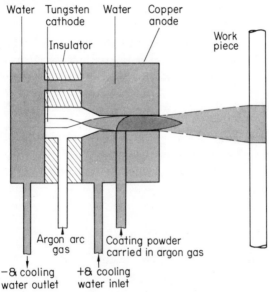

Fig. 3. The plasma gun (schematic).

The very high temperatures in the plasma torch allow any material to be sprayed, provided it melts without significant dissociation and provided a practical temperature interval exists between its melting and boiling points; the absence of direct substrate heating minimises the problem of workpiece distortion and allows the use of low, as well as high, melting point bases. The process is thus remarkably versatile; it may be used to spray metals and alloys, ceramics and cermets and plastics, while the substrate may be of metal, ceramic or plastic.

15

Plasma deposits can vary in thickness from less than 0·03 mm to more than 2·5 mm; normally, however, they are in the 0·10–0·60 mm range. Due to the high particle temperatures and the rapid impingement on the target, the coatings are usually dense and adherent—porosities as low as 1% being common—while the presence of an inert atmosphere during spraying minimises oxidation of the coating. Within limits, the properties of the deposit can be varied to suit particular applications, e.g. a controlled degree of porosity may be incorporated to enhance oil retention for bearing applications.

Plasma coatings are being increasingly used for wear and abrasion resistance, electrical conduction, thermal and electrical resistance, reclamation of worn parts, repair of wrongly machined components and, in certain cases, corrosion resistance. Fig. 4 shows a few examples of coated articles. In addition to surface treatment applications, the process is also finding growing use for the production of free-standing shapes and special powders; in these cases, steps are taken to prevent the deposit adhering firmly to the substrate.

Fig. 4. Selection of small, plasma sprayed components.

Table 2. Features of thermal spraying processes

	FLAME	ARC	PLASMA	DETONATION
Impingement speed, m/sec.	100	100	150	800
Approx. Temperature, K	3,000	5,000	20,000	4,000
Typical deposit porosity, %	10–15	10–15	1–10	1–2
Typical adherence, MN/m^2	7	10	30	60
Advantages	Cheap. High deposition rates.	Cheap. Low contamination. Cool substrate.	Low porosity. Good adhesion. Versatile. Cool substrate. Low contamination.	Very low porosity. Very good adhesion. Cool substrate.
Limitations	Normally high porosity. Inferior adhesion. Heats workpiece. Limited.	Only electrical conductors. Normally high porosity.	Fairly expensive.	Very expensive. Low efficiency.

17

Modern spraying practice has tended to centre around the four methods listed earlier. Each has its advantages and limitations (see Table 2) so that distinct (but overlapping) areas of applications are developing. In general, the quality of the coating and the cost of spraying increase in the order: flame, arc, plasma, detonation. In all cases, however, satisfactory results can only be obtained if care is taken to optimise spraying conditions and especially substrate preparation; at the same time, good deposits can be improved by subsequent treatment.

2. The Physics of Plasma Spraying

The injection of a stream of powder particles via a carrier gas into a plasma jet, so that it is deposited as molten droplets on the workpiece, is associated with a wide range of variables. To optimise the plasma spraying process it is necessary to understand the factors affecting arc-particle interactions and the techniques by which they may be controlled.

2.1. THE PLASMA ARC

In an arc, free burning between two electrodes, electrons emitted from the cathode are accelerated by the applied potential. The electrons acquire high kinetic energy and collide with atoms and molecules of gas. This causes ionisation of the atoms and dissociation in diatomic gases; recombination subsequently takes place, both heat and light being evolved. In this way, the electrical energy applied between the electrodes is converted to heat and the temperature of the arc is 5,000–6,000 K. The thermal efficiency depends upon many factors but lies between 30 and 70%.

In free burning arcs, the heated gas expands laterally and this decreases the enthalpy (or heat content). Increase in power increases the volume of plasma but only raises the arc temperature slightly; to achieve significantly higher temperatures, it is necessary to increase the current density and the enthalpy. This—an essential feature of the plasma torch—is usually done by passing the arc through an anode nozzle of diameter smaller than that of the natural arc (Fig. 5). It may be noted, incidentally, that commercial plasma guns contain only partially ionised gases so that they are not strictly true plasmas.

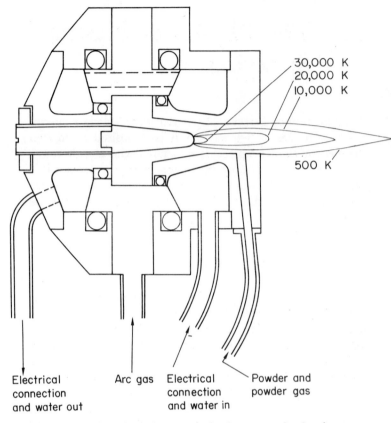

Fig. 5. Section through a typical plasma torch showing constricting nozzle and temperature distribution.

In a constricted arc, the primary heat transfer process is thermal conduction and large radial temperature gradients exist. A 5 mm diameter arc may have a core temperature of more than 20,000 K but a temperature at the nozzle wall of only 500 K (see Fig. 5). These large gradients naturally affect the spraying process. The relationship between arc radius and voltage gradient is given by the Ellenbaas–Heller equation, which is the arc energy balance; neglecting radiation:

$$\sigma E^2 = -\frac{1}{r}\frac{\mathrm{d}}{\mathrm{d}r}\left(rK\frac{\mathrm{d}T}{\mathrm{d}r}\right)$$

where σ is the electrical conductivity,
$\quad\quad E$ is the voltage gradient,
$\quad\quad r$ is the radius,
$\quad\quad K$ is the coefficient of thermal conductivity,
and $\quad T$ is the absolute temperature.

The term on the left is the electrical power input to an elementary cylindrical shell of unit volume and that on the right the radial heat flow through this shell due to the temperature gradient.

Increase in voltage gradient means that power input per unit length of arc, P, is increased and the core temperature is raised in proportion to $P^{2/7}$ (as shown by L. A. King). In the plasma gun, an axial flow is passed through the nozzle and this cools the arc boundary; solution of the Ellenbaas–Heller equation for such conditions of increased radial heat flow indicates a further increase in voltage gradient and decrease in arc radius. This causes an increase in central core temperature, which is dependent on the rate of flow of gas or rate of cooling. Thus an increase in peripheral arc cooling has the surprising effect of raising the core temperature, provided the current remains constant.

2.2. THERMAL CONDUCTIVITY AND HEAT TRANSFER

Thermal conductivity of the arc gas and the heat transfer mechanism within the arc are of prime importance in plasma spraying. In the presence of high temperature gradients, the classical conductivity of gases at high temperatures is enhanced by a process of radial diffusion of atoms and ions from the high temperature regions, in which they are formed, to lower temperature recombination zones.

The variation in thermal conductivity of a diatomic gas with increase in temperature is shown in Fig. 6; the various recombination processes give an additive liberation of heat and such processes are more likely to occur on the particle surface. Heat transfer to particles is thus extremely efficient and far exceeds that expected from classical considerations.

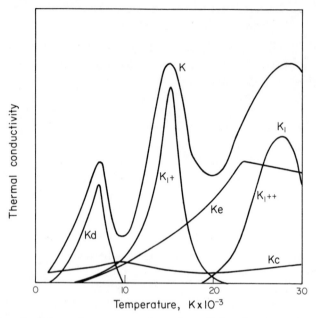

Fig. 6. Schematic representation of variation in thermal conductivity of a diatomic gas with temperature.

The large temperature gradients in the arc, coupled with the variable geometry of the particles and the consequently varying trajectories, make an exact analysis of the heat transfer situation difficult and various approximate treatments have been developed. An empirical approach has been to calculate the largest diameter of particle which would attain, at its centre, a temperature of 90% of its melting point with a gun dwell time of 100 μsec; this is known as the "difficulty of melting" parameter. Essentially, the smaller this diameter the more difficult the material will be to melt. This treatment reduces the problem to one of conduction.

A more analytical approach expresses certain basic material parameters (specific heat, latent heat of fusion, melting point and density) as a single factor, L, the heat content per unit volume of the liquid metal at its melting point, relative to that of the solid metal at room temperature. The difficulty of melting is then proportional to $L\rho^{-\frac{1}{2}}$ where ρ is the particle density. The heat transfer from plasma to particle is described by the Engelke equation:

$$\frac{S(\lambda\Delta T)^2}{V\mu}\text{ plasma} = \frac{L^2 D^2}{16\rho}\text{particle}$$

where S = mean plasma length
V = mean plasma velocity
λ = mean thermal conductivity of the boundary layer between gas and particle
T = mean temperature drop in the boundary layer
μ = mean plasma viscosity
L = heat content of unit volume of a particle in liquid condition at the melting point relative to that of the solid at 293 K
D = average particle diameter
ρ = average particle density

Table 3 shows "difficulty of melting" parameters for a range of materials.

Table 3. $L/\sqrt{\rho}$ Parameters for various materials (from K. Hanush, R. Kammel and H. Winte-hager, Blech, 1968, (8), pp 452–61)

MATERIAL	MELTING POINT, K	DENSITY, gm/cc (ρ)	HEAT CONTENT AT MELTING POINT (L)	$L/\sqrt{\rho}$
TiC	3,410	4·93	4·72	2·13
TiN	3,200	5·21	4·56	2·00
ZrC	3,840	6·73	4·48	1·73
TaC	4,150	14·53	4·99	1·31
Al$_2$O$_3$	2,318	3·97	3·39	1·70
W	3,650	19·30	3·45	0·79
Ta	3,270	16·60	2·59	0·64
Ni	1,728	8·90	2·37	0·79

2.3. PARTICLE TRAJECTORY AND DWELL TIME

Within the plasma jet there are concentric zones of varying thermal conductivity and radially reducing temperature. Particles fired radially into the arc pass through these zones and receive an amount of heat energy which depends on trajectory and dwell time in the arc. For the most efficient heating, it is not necessary for a powder particle to pass through the arc core since it could conceivably be heated to a higher temperature by passing longitudinally through one of the recombination zones. However, to attain an approach to the ideal, in which all particles are melted but few evaporate, similar trajectories are required for all. Since trajectories are dependent on gas flow and particle size, successful plasma spraying requires a close size-range of powder. Fig. 7, which is an enlarged section of Fig. 5, demonstrates these effects.

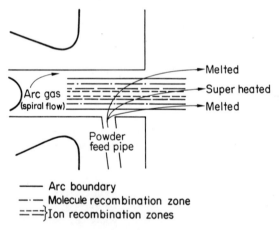

Fig. 7. Section through the plasma gun nozzle showing particle trajectories and recombination zones.

High-speed photography has shown that considerable fluctuations in arc length occur in the plasma jet (Fig. 8) and these have been attributed to the action of the gas in blowing the arc into a "hairpin" configuration. The loop of the hairpin is shorted out when the voltage drop around the loop is equal to the breakdown voltage between the anode root and the column of the arc. The process is then repeated. Such effects occur up to a critical current (50 A for argon and 200 A for nitrogen), above which the anode root mechanism changes and a number of independently fluctuating hairpins are produced; at the currents commonly used in plasma spraying the phenomenon is most significant with nitrogen. A fluctuating single hairpin is undesirable in practice since it results in large variations in arc dwell time.

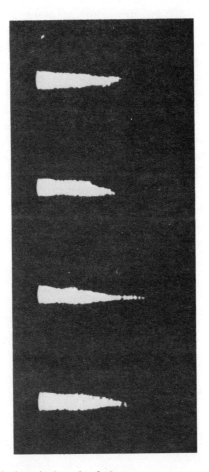

Fig. 8. Variations in length of plasma arc.

For a given gas flow rate, smaller particles attain a much higher velocity than larger particles. High velocity is associated with low contamination and good deposit characteristics; there is, consequently, considerable interest in spraying fine powders in special high velocity guns, from which efflux velocities close to sonic speeds can be achieved.

2.4. ARC GAS

The choice of arc gas in plasma spraying is dictated by chemical, electro-thermal and economic considerations; both monatomic and diatomic gases are used industrially and these have different specific heat and dissociation energy requirements. As Fig. 9 shows, the diatomic gases have higher specific enthalpies than the monatomic. However, they also have higher effective thermal conductivities and hence the arcs have a smaller volume so that, although the peak temperature of a diatomic gas arc may be higher, the deposition efficiency tends to be lower than with a monatomic gas. The electric power/gas heating efficiency for all the usual gases appears to be of the order of 60%.

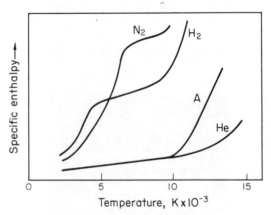

Fig. 9. Specific enthalpy–temperature relationships for gases used in plasma spraying.

3. Plasma Spraying Equipment

A plasma spraying unit consists essentially of a plasma torch, supplies and services, a control system and ancillary equipment (Figs. 10 and 11). Each of the equipment manufacturers listed in the Appendix incorporates a number of special features, particularly in gun design. It is not intended, in the limited space of this chapter, to

Fig. 10. Plasma spraying equipment.

describe all these variants but rather, by reference to typical equipment, to illustrate the salient features of commercial spraying sets.

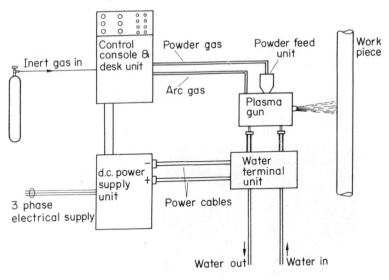

Fig. 11. Typical plasma system.

3.1. PLASMA TORCHES

Typical plasma torches are shown in Figs. 12 and 13 and, in exploded view, in Fig. 14. These guns operate in the non-transferred mode, in which the arc is struck within the torch. In the transferred arc system, the external workpiece acts as the anode (Fig. 15); however, because the substrate receives a considerable amount of heat, this arrangement finds little use in spraying practice.

In a typical gun, there is a front nozzle-shaped anode and a rear pointed cathode. To withstand the intense heat, the electrodes are water-cooled. Copper, because of its high thermal conductivity, is the usual anode material while thoriated tungsten, which is a good electron emitter, is usually chosen for the cathode. The carrier gas flows around the annulus between the electrodes, forcing the arc into the nozzle and so constricting it. The arc is stabilised to prevent it impinging on, and rapidly eroding, the anode; in the widely used vortex method, the carrier gas is injected tangentially and sets up a swirling action. The low pressure near the centre of the vortex expands the arc and stabilises it as it passes through the nozzle. The electrodes are arranged concentrically to prevent the arc being driven too close to either wall. Flow in the nozzle is laminar but becomes turbulent at the nozzle exit and this facilitates transfer of energy from the arc gas.

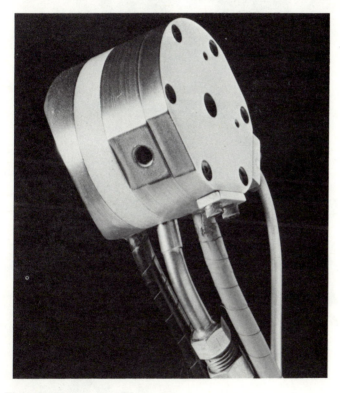

Fig. 12. Plasma spraying torch.

Plasma torch geometry has a marked effect on spraying efficiency. The nozzle diameter is a particularly important variable, since changes in it affect the arc gas velocity, the current density, the powder velocity and particle trajectory; in general, increasing the nozzle diameter

c

29

Fig. 13. Plasma spraying torch (with cover).

Fig. 14. Exploded view of a typical plasma torch.

increases the spraying efficiency but weakens tne adhesion of the deposit. It is necessary, however, to select the nozzle to suit the gas and powder to be used and no single nozzle will give optimum performance over the range. Nozzle cooling is also important since it affects not only the degree of erosion-resistance of the nozzle but also the power capability of the set. Finally, the length of the

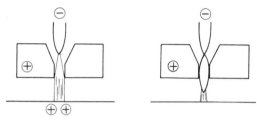

(a) Transferred mode   (b) Non-transferred mode

Fig. 15. Plasma spraying modes.

Fig. 16. Automatically controlled gun and workpiece handling equipment.

nozzle influences the stability and direction of the arc. To spray the more refractory materials, guns have been developed with a short cathode and a long, small diameter nozzle anode; this arrangement leads to high velocity gas flows and very high efflux velocities. Coupled with the use of argon/hydrogen gas mixtures, increased temperatures and impingement velocities on the substrate are achieved and these give best quality deposits of materials such as tungsten carbide/cobalt.

The plasma gun may be hand-held—as in traditional flame spraying—or it may be controlled by a suitable rastering device (Fig. 16). The former gives a less uniform deposit than the latter, so that the highest quality coatings are generally obtained with automatic control of the torch. The ability to spray internal surfaces is limited to bores which may be penetrated by the gun (except where the component is shallow). For this reason miniature guns have been developed (Fig. 17); some are claimed to be able to spray bores of as little as 50 mm. These guns, however, are of rather limited power capacity.

Fig. 17. Inside surface of a tube being sprayed with a "miniature" plasma gun.

3.2 ARC POWER

Most plasma spraying sets operate on d.c. power. Although a.c. appears to have some economic advantages, especially at high power levels, it has given problems with arc instability, high nozzle erosion and low conversion efficiency and these have, so far, precluded its use.

Welding-type rectifiers, with a drooping characteristic to prevent instability of the arc, are generally used. These require a high open circuit voltage to cover the range of gases used; typical voltages are 80 V for argon, 160 V for nitrogen and 300 V for hydrogen. Arc voltages range from 20 to 80 V and typical spraying units have a power rating of 20–40 kW. It is normal to initiate the arc with a high frequency discharge.

3.3 ARC GASES

The gas introduced into the plasma torch must sustain the arc and transport the powder feed. The choice of gas, however, affects electrode life, spraying efficiency and deposit quality. Nitrogen, hydrogen, argon and helium are the four gases that are used commercially; their characteristics are summarised in Table 4. Since, for a given material, the deposition efficiency has an optimum value at a given arc enthalpy, the correct gas should be selected for a particular material. Irrespective of the gas used, its quality affects electrode life; in particular, moisture and oxygen levels must be low.

Nitrogen, the cheapest of the gases, has been widely used and possesses good heat transfer characteristics. However, it can give problems with arc length (as mentioned in 2.3), with electrode erosion and with nitrogen absorption in the coating; it is mainly used to spray stable oxides and to achieve high deposition rates.

Argon is, in general, the most suitable for routine spraying; efficiencies normally exceed 75% and can exceed 90%.

Table 4. Data on arc gases

GAS	VALENCY	RELATIVE HEAT CONTENT	PER CENT USED	REACTIVITY	RELATIVE NOZZLE LIFE	COST	REMARKS
Nitrogen	Diatomic	High	>80	May react to give nitride	Poor	Low	High deposition rates, good heat transfer
Argon	Monatomic	Low	>75	Inert	Good	Moderate	Best general purpose gas, high efficiency
Hydrogen	Diatomic	High	5–25	Reducing	Poor if too high H_2 content	Moderate	Hot flame, high expansion, used for refractory metals such as WC/Co
Helium	Monatomic	Low	10–20	Inert	Good	High	Hot flame, used to avoid H_2 embrittlement

The high cost of the gas, compared with nitrogen, is more than offset by the saving in powder consumption. Argon gives the longest electrode life and does not react with sprayed deposits.

Neither hydrogen nor helium is usually used as a primary gas, but either may be added to argon or nitrogen, generally to increase the heat transfer properties; the main applications are, therefore, when spraying refractory materials. As an example, it has been found that tungsten carbide/cobalt may be sprayed with increased efficiency by the addition of 10% hydrogen to the argon arc gas. The hydrogen addition results in a hotter and more concentrated plasma; by using the high velocity type of gun, described in an earlier section, these very high melting point materials may be deposited with maximum density. Small hydrogen additions may be used to obtain reducing conditions for the deposition of readily oxidised materials.

Air normally becomes entrained in the carrier gun as it leaves the nozzle and the concentration of the protective gas may be substantially reduced before the particles hit the substrate. Due to the short dwell time of particles in the plasma process, this normally gives little trouble. However, where oxidation does prove a problem, techniques have been developed to shroud the nozzle, or, alternatively, to spray in an inert gas chamber. The latter poses many problems and is rarely used. However, by fitting a water-cooled conical shroud to the front of the gun and extending it to the surface being sprayed, satisfactory results can be obtained (see Figs. 18 and 19).

3.4. POWDERS

Although a few early plasma spraying guns were designed to operate on a wire feed, all commercial torches use powders. A compromise may be necessary between a closely sized powder, which gives easiest sprayability, and a wider range powder, which gives minimum porosity coatings; in all cases, the powder should be dry and fairly free-flowing.

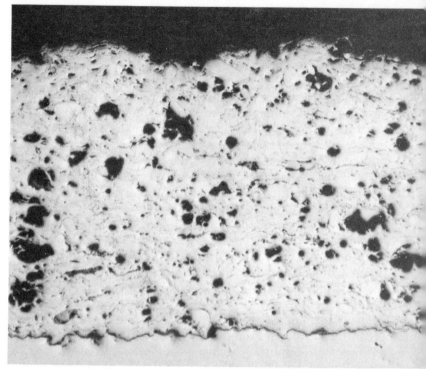

Fig. 18. Plasma deposited chromium, sprayed without shroud. (×75.)

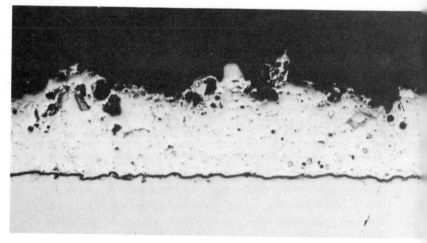

Fig. 19. Plasma deposited chromium, sprayed with shroud. (×75).

The powder feed system must be able to deliver accurately metered quantities of powder through long flexible lines with accurately controlled amounts of carrier gas; it must also be possible to vary the powder and gas flow and to handle a variety of powder densities and particle sizes. Various types of feed mechanism, based on the aspirator or the mechanical metering type, are commercially used. Fig. 20 shows a hopper design that has proved satisfactory. The gap between the inverted cone and the tube through which the powder falls can be adjusted by a calibrated screw thread. The hopper is pressurised and the gas passes into a swirl chamber to collect powder falling past the cone and passes out through inclined radial holes. The hopper is vibrated by an air turbulator operating at a controlled variable pressure. It is suitable for powder particles down to ~400 mesh size.

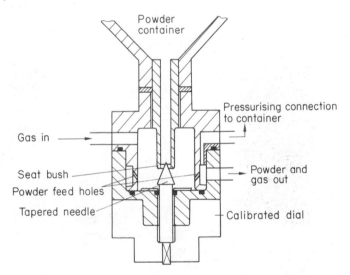

Fig. 20. Power feed unit.

The gas flow rate and powder feed rate, and the point and direction of entry of the powder into the arc, affect the spraying efficiency and/or coating quality. For high melting point materials, it is desirable to have a long dwell time in the arc and the powder feed is, therefore, located as far back along the nozzle as possible; with plastics, a short dwell time is sufficient and the powder is introduced at various points outside the nozzle.

3.5. CONTROL FACILITIES

The control facilities may range from manually operated sets with few controls to comprehensive installations that continuously monitor the process and can be fully automated. Fig. 21 shows a console which is fully automatic and contains provision for controlling current, voltage, 3 arc gases, 2 powder gases, 2 air-powered vibrators for powder hoppers, cooling and deflector jets, in addition to the power supply, water pump and extraction controls and the safety interlock. Since the workpiece and gun are automatically controlled, operation of the set is possible from outside the spraying chamber.

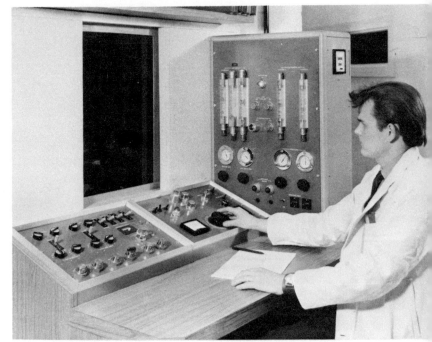

Fig. 21. Plasma console with extensive range of controls.

3.6. ANCILLARY EQUIPMENT

The substrate (or workpiece) must be completely free from dust, oil, etc., and for this reason degreasing and grit-blasting facilities are necessary. A vapour degreasing tank is usual and, for grit blasting, a standard machine is generally satisfactory; masks or masking tape are used to protect areas within the field of blasting.

The workpiece is normally located, during spraying, on a lathe or similar equipment. In most sets, the spraying torch is hand held but automatic control of the gun, along an X–Y raster grid, is also used; a typical assembly is shown in Fig. 16. This not only allows external control of the process, with consequently safer operation, but gives more uniform coatings.

Most sprayed surfaces are finished by a machining or grinding step; in the case of very hard deposits (such as tungsten carbide), diamond grinding facilities are essential.

3.7. HEALTH HAZARDS

Plasma spraying units provide certain potential health hazards but, with sensible operation and safety precautions, these become minimal.

(i) Power sources are similar to those of welding generators and the same precautions should be observed.

(ii) The very high energy outputs require substantial cooling water (e.g. 0.8 m^3/hr at a delivery pressure of 0.55 MN/m^2).

(iii) The arc plasma radiates intensely at certain wavelengths and the ultraviolet radiation is very severe. The eyes and skin surfaces should be protected by suitable clothing and shielding.

(iv) Adequate ventilation is necessary to remove ozone, which is generated by the ultraviolet radiation, and any oxides of nitrogen. Large quantities of metal or ceramic dust are given off during spraying and these are extremely hazardous. With hand spraying, protective breathing apparatus is used, but the safer method is to have completely automatic control and operate the set from outside the spraying booth.

4. The Practice of Plasma Spraying

As with other surface treatments, the successful application of plasma spraying depends to a very high degree on the prior surface preparation of the substrate. With inadequate preparation there can be no successful application. The steps necessary for plasma spraying are listed in Table 5.

Table 5. Steps in successful plasma spraying process

1	Clean	Remove paint, corrosion products, dirt, grease.
2	Grit blast	To finish cleaning and provide rough, high activity surface.
3	Spray	Deposit coating.
4	Post process	Seal, infiltrate, sinter.
5	Finish machine	Buff, turn, mill, grind.

Note: Not all steps are necessary in all cases (e.g. stage 4 is used only for special applications).

4.1. COMPONENT DESIGN

The design of a part submitted for plasma spraying is the first consideration. If it is possible to place the gun so that the anode nozzle is 50–75 mm from the workpiece, with the gun axis normal to the surface to be coated, then the part is likely to be amenable to plasma spraying. For narrow bores into which the gun cannot enter, it is possible to angle the gun to spray from outside the bore, but in this way it is only possible to spray narrow regions, since variations from the normal spraying angle of greater than 15° result in reduced adhesion and increased porosity. Sharp corners at the edge of a surface to be sprayed should be chamfered, of the order of 0·1 mm from the apex, to prevent any tendency for the deposit to lift at the edges. Grit blasting rounds-off such sharp corners, and this may be sufficient. If it is desired to spray into a recessed groove, the walls of the groove should be angled at a minimum of 60° to the spraying axis. This avoids the problem of shadowing, which occurs particularly with gun-axis to groove-wall angles less than 45°, whereby gross porosity and poor bonding can be present at the wall root.

In addition to the surface features of the part to be sprayed, the nature and thermal capacity of the substrate must be considered. For spraying onto most metal substrates heat input is not significant in the plasma spraying process since the part is unlikely to rise above 373 K with normal practice. However, with delicate, close tolerance, low thermal capacity work, and especially ceramic, glass and plastic substrates, further attention must be given to the acceptable level of heat input. Deposition rate, traverse speed and cooling gas flow should be adjusted for very low heat input. The thermal capacity of the molten droplets should be considered and the greatest care taken with high melting point, high specific heat materials. The use of the monatomic arc gases is to be preferred since their lower specific enthalpy as compared with diatomic gases should result in less workpiece heating. The thickness and nature of the sprayed deposit should be specified, consistent with end use, to minimise thermal stresses.

4.2. SUBSTRATE PREPARATION

To obtain optimum adhesion, the surface of the part to be sprayed must be clean and free from grease. Paint and loose rust are removed by pickling, scraping or wire-brushing. Vapour degreasing, or for the largest parts, liquid degreasing, is then performed in trichloroethylene. Following degreasing, handling of the workpiece should be minimised. Where it is necessary to touch the workpiece near the areas to be sprayed, the use of clean cotton gloves is recommended.

Since the bond in a plasma sprayed deposit is predominantly mechanical, the surface to be sprayed is roughened by grit blasting (Fig. 22); this produces a rough, clean, high activity surface which does not require a thick coating to fill. Areas to be protected from grit blasting, machined surfaces and areas where coating is not required, are protected by the use of metal masks or PVC adhesive tape. The grit-blasting air should be clean, dry and oil-free. High gas pressures, of the order of 0.6 MN/m^2, are to be recommended to minimise blasting time. At such pressure, grit life is short but the saving in labour costs more than compensates for this. For delicate components, lower pressures, down to 0.2 MN/m^2, should be used. The grit should be clean, fresh and sharp, containing a minimum

Fig. 22. Mild steel surface, grit blasted with 20-mesh white alumina. ($\times 120$).

43

of fines; chilled cast iron and aluminia grits are widely used. Rusty or dirty components are generally blasted in two stages using old grit first, followed by fresh grit. Following grit blasting the part is cleaned in an air blast. Masking unsuitable for plasma spraying is removed. To minimise contamination, handling of the part should be reduced and spraying should preferably be performed within one hour of grit blasting.

4.3 PLASMA SPRAYING

Prior to spraying the component, it is necessary to protect those areas where overspray is not permissible. Protective paint, PVC tape protected by a metal shield or heat-resistant adhesive tape are used for this purpose. The workpiece is set up in the handling equipment, a turn-table, lathe or clamp, and the gun traverse limits are set. Cooling jets are arranged so that they impinge on the workpiece but do not interfere with the spraying process. For thin material, jets may be positioned at the back of the workpiece opposite the sprayed area, while deflector jets may be set up to divert the hot gases from the sub-strate. The gas flow of the deflector jets must be low so the path of the sprayed particles is unaffected. Hand spraying may be necessary on parts of complex geometry, but the high degree of control achieved by automatic handling favours its use wherever possible.

If the deposition parameters have not been established, an efficiency test is performed by the method outlined in Chapter 5; if the deposition parameters are known, spray-ing may begin under the established conditions. The gun normally traverses a distance greater than the length of the sprayed area so that end effects are not apparent on the workpiece. After the calculated number of passes, the workpiece geometry is checked to determine whether further passes are required. When the specified coating thickness has been attained, the part is removed from the handling equipment and inspected. Prior to the finishing operations, the part is cleaned and the protective tape and paint are removed.

Apart from the nature of the arc/powder gas and the powder itself, the variables in the plasma spraying process include power input, spraying distance and angle, arc and powder gas flow rates and hopper setting. Table 6 lists optimal settings for a range of powders. Increasing hopper setting results in increased feed rate, but too high a setting results in a reduction of efficiency. The powder gas flow rate controls the trajectory of the particles through the arc and must be balanced with the arc gas flow rate. Increase in arc gas flow rate results in decrease in arc diameter, increase in arc temperature and specific enthalpy, and increase in particle velocity. Power input is determined to a large degree by the nature of the gas, but increase in power results in increased arc temperatures. Spraying distance is an important parameter in its effect on substrate heating and particle dwell time.

4.4. POST PROCESSING

For many applications, porosity (with consequent permeability and inferior strength) is undesirable; as an example, the use of plasma sprayed deposits for chemical protection is effective only if the attacking media are totally excluded. The painting of a fluid varnish on the surface of a plasma sprayed deposit is one technique by which this can be achieved. The varnish, chosen for the specific application from the range of proprietary compositions, is drawn into the pores by capillary action. Such a technique provides fully dense surfaces for applications at temperatures up to about 400 K; for higher temperatures up to about 600 K, the use of a high strength adhesive is recommended. These adhesives also provide up to five-fold bond and cohesive strength reinforcement. Epoxy resin adhesives, with their high curing fluidity and high strength, have proved eminently suited to this type of application. For even higher temperature applications, infiltration with molten metal similar to that practised in powder metallurgy can be performed; for example, copper readily infiltrates porous ferrous bodies. Densification can also be achieved by sintering and there is evidence that

D

Table 6. Standard spraying conditions

MATERIAL	ARC GAS FLOW RATE, m³/hr	POWDER GAS FLOW RATE, m³/hr	DISTANCE mm	HOPPER SETTING	FEED RATE, g/min	EFFICIENCY, %	REMARKS
Alumina	1·0	0·34	40–50	11	20	85	
Aluminium	1·5	0·27	75	8	19	95	
Al bronze	1·4	0·29	75	10	64	85	
Chromium	1·4	0·31	50–75	10	78	87	
Chromic oxide	0·8	0·24	50	10	34	80	
Copper	1·5	0·31	75	8	73	77	
Molybdenum	1·4	0·24	75	6	20	88	
Nickel	1·4	0·27	75	6	75	80	
Silicon	1·4	0·31	50	2	16	51	Porous
Stainless steel	1·4	0·31	75	8	39	73	
Titanium	1·5	0·24	50	6	23	90	Argon shroud
Tungsten Carbide/ Cobalt	1·5	0·34	50	2	31	72	High velocity gun, argon + 10% hydrogen
Zirconia	1·5	0·31	40	8	25	64	

Arc gas : argon
Current : 500 amps

strengthening of plasma sprayed materials occurs at much lower temperatures than those used in the sintering of conventionally pressed powders.

The above range of techniques can be equally applied to both plasma sprayed surfaces and free-standing shapes. The latter are prepared by spraying onto a smooth, shaped mandrel, usually manufactured in copper, which has been aerosol sprayed prior to plasma spraying with a salt solution which acts as a release agent. When the required thickness has been built up, the mandrel is immersed in a stream of water which dissolves the salt and releases the free-standing shape. Using this technique, materials which are difficult or impossible to process by any other means can be readily formed. Densification by one of the above techniques can be performed to produce a strong, dense body.

4.5. FINISH MACHINING

Where a close tolerance or smooth finish is required, the plasma sprayed surface must be machined. The control of the plasma spraying process, using the automatic equipment described above, is such that a coating can be deposited under normal conditions to within 0·02 mm of a specified thickness with a surface finish of 70 μm C.L.A., which is acceptable for many applications. Using specialised techniques involving ultrafine powders, these values can be reduced. Machining of coatings is usually involved with the removal of 0·1–0·3 mm, although coatings are usually smooth after the removal of 0·1 mm, and the removal of more than this is wasteful.

The softest coatings, e.g. plastic, plastic/metal and lead, can be buffed smooth. The softer metals, aluminium, aluminium-bronze, pure iron and copper and plastics can be machined with conventional cutting tools. Metals such as chromium, molybdenum, stainless steel and tungsten require grinding, and tungsten carbide/cobalt and most ceramics must be diamond ground.

5. Deposit Testing and Quality Control

The testing of deposits is an essential part of any coating development and must be a routine feature of all quality control. Although the application is the best test facility, it is generally necessary to use laboratory tests for assessment purposes. Some of these are well established and yield reliable results; in others (e.g. the measurement of bond strength) the results must be interpreted with caution. The principal tests used for coatings are briefly described in this chapter. In addition to these routine tests, more specialised tests—chosen according to the component application—are often used to evaluate, for example, corrosion resistance, wear rate, thermal stability or thermal shock resistance, while the assessment may include service simulation tests.

5.1 POWDER EVALUATION

Successful plasma deposition depends upon the use of the correct feed material, and powder evaluation is widely practised; normal powder metallurgy tests are suitable for this purpose. Chemical analysis, powder size, shape and area, powder density and flow rate are generally measured.

The powder samples must be representative and the method of sampling is prescribed. Standard chemical analysis techniques are used, but it is usual to include hydrogen loss and oxygen level. The apparent density of the powder can be measured in a flowmeter and calibrated cup, the weight of dry powder to fill the cup being used to determine density. The time for a standard amount of powder to pass through the flowmeter gives the flow rate.

The particle size characteristics are measured by normal sieving, using automatic agitation of the sieves; methods for sub-sieve powders are still under development. Particle shape is assessed by examining the powder under a low-power microscope, the shape being reported as acicular, irregular, spherical, etc. Surface area can be determined by surface adsorption of nitrogen.

5.2. DEPOSITION EFFICIENCY

The efficiency of deposition is taken as the ratio of the weight of material deposited to the weight of powder used, expressed as a percentage. Fig. 23 shows an experimental arrangement by which the deposition efficiency of a large range of commercial powders has been measured.

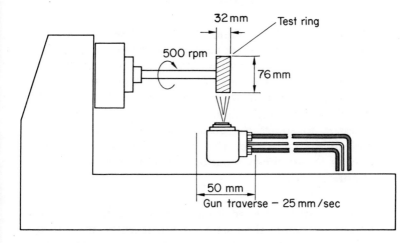

Fig. 23. Test arrangement for establishing optimum spraying conditions.

A test ring, suitably prepared and weighed, is mounted on a mandrel in a lathe chuck. The gun, mounted on the lathe saddle, is set to traverse across the face of the ring with an overspray. Spraying is performed for exactly one minute. The powder feed rate is given by the weight of powder used (P). The deposition rate is the weight increase of the ring (R), multiplied by total traverse width (T), divided by ring width (W). Thus deposition efficiency is given by:

$$\frac{R \times T}{W \times P} \times 100\%$$

The test may be carried out in replicate. Using such deposition tests, the spraying parameters can be optimised for any given powder; however, it is also necessary to check coating structure, bond quality and porosity level (by metallographic examination) to ensure that spraying efficiency is not improved at the expense of deposit quality.

5.3. METALLOGRAPHIC EXAMINATION AND POROSITY MEASUREMENT

Metallographic examination of polished sections of the deposit and substrate plays a major part in the evaluation procedure for general coating quality and integrity, porosity, inclusion content, etc. The methods used to prepare specimens are not, in themselves, different from normal metallographic preparation techniques, but great care is necessary to ensure the absence of artefacts that cause misleading results. For this reason, very detailed procedures for specimen mounting, grinding and polishing are prescribed.

Porosity can be assessed from examination of a metallographic section of the coating. Until recently, this led to difficulties in interpretation but the introduction of quantitative metallographic techniques has given a rapid and reliable method of measurement, which is rapidly becoming standard. Where such facilities are not available, the interconnected porosity may be measured by an oil immersion technique, while the overall density of the deposit can be obtained from a standard powder density determination on a coating removed from the substrate; neither method, however, gives very accurate results.

5.4. MECHANICAL PROPERTIES

The adhesion (or bond strength) of the coating/substrate interface and the hardness of the deposit are the mechanical properties most usually measured; the tensile strength of the coating may also be determined. Care must be taken with test design and procedure if reliable and reproducible results are to be obtained.

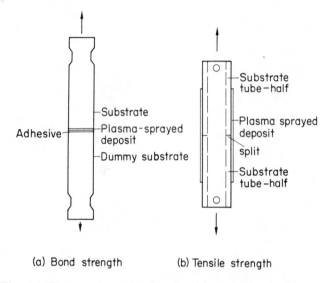

(a) Bond strength   (b) Tensile strength

Fig. 24. Test arrangements for plasma sprayed coatings.

The adhesion test shown in Fig. 24 is widely used as an acceptance test for sprayed coatings, the halves of the specimen being bonded with a suitable resin, normally an epoxy. Despite its extensive use, discrepancies in the results have led to recent re-examination of the technique and it has been shown that, with epoxy resins and normal bonding practice, significant penetration of the resin to the coating/substrate interface takes place, and this can give erroneously high results. When suitable precautions are taken, by sealing the coating to prevent penetration of epoxy or by the use of a more viscous resin, this can be prevented and a more reliable adhesion figure is obtained; Table 7 compares results using the usual and the improved bonding method. It is possible to determine whether the test has been valid by examining the fracture. Separation at the coating/substrate boundary, or failure wholly within the coating, indicates a reliable test; a mixed fracture, partly in the deposit and partly in the bulk material, is considered invalid (see Fig. 25).

Fig. 25. Adhesion samples after testing. Left pair: mixed fracture; Right pair: coating/substrate separation.

A simple and fairly reliable test for coating strength is based upon two tubes, of the same diameter and accurately machined to give mating faces (Fig. 24). The tubes are mounted on a mandrel and sprayed along their length. The ends of the tubes are threaded to fit a tensile testing machine and, after removal of the mandrel, the coating may be pulled to failure.

In macro-hardness testing, the coating must be greater than a minimum thickness (related to the test load) to obtain reliable results; otherwise penetration of the indentor or interference from the substrate will yield spurious results. The relatively large ball of the Brinell indentor is more suitable for measuring the bulk hardness of coatings than the smaller indentor of the Vickers or Rockwell hardness testers. The hardness of the individual particles in a coating is best measured by micro-hardness indentations on transverse sections of the particle.

Table 7. Adhesion of plasma deposits with different bonding agents

BONDING AGENT	SPRAYED MATERIAL	AVERAGE BOND STRENGTH, MN/m^2
A	Aluminium Bronze	44·8
	Molybdenum	51·7
	Zirconia	39·3
B	Aluminium Bronze	21·0
	Molybdenum	22·4
	Zirconia	38·2

All tests were on 20·2 mm diameter aluminium specimens with 0·125 mm coatings.

    Bonding Agent A: Usual epoxy
    Bonding Agent B: More viscous resin

5.5. QUALITY CONTROL

Quality control of deposits is a very important parameter since so many plasma coatings find use in high integrity applications. Faulty materials or engineering, or incorrect practice, can produce a coating of low adhesion, the existence of which may be unperceived until failure occurs in service. The situation is particularly difficult because of the lack of satisfactory non-destructive test methods. The current practice is to perform tests (e.g. adhesion tests) on specimens sprayed under nominally the same conditions as the workpiece. Efficiency checks performed before and after spraying are used to ensure consistency of deposition conditions. The microstructure of the deposit on the efficiency test ring will also be examined for a visual check of coating and bond quality. Despite these tests, quality assurance in sprayed deposits relies heavily on the observance of good practice at all stages in the process.

6. Deposit Characteristics

One of the advantages of the plasma process is its ability to produce deposits whose properties can be varied, within limits, to meet specific requirements. This versatility presents difficulties when attempts are made to compare the reported properties of coatings and these are increased when different testing conditions have been used. The results quoted in this chapter should, therefore, be regarded only as a general guide.

6.1. MATERIALS SELECTION

Although there is an almost limitless number of material combinations that may be plasma sprayed, commercial practice is largely based on about 20 or 30 (See Table 8). For special applications, however, other materials may be chosen to give special properties in the finished deposit (e.g. self-lubrication, low thermal expansion, etc.).

6.2 SPRAYING EFFICIENCY

The measurement of spraying efficiency has already been described and representative values for different powders have been listed in Table 6. Although this is a parameter of considerable interest, it should be repeated that, in practice, it may be necessary to depart substantially from optimum spraying conditions to obtain the required properties in the deposit. For the same reason, quoted deposition rates for different materials are of little significance unless related to the deposit application.

6.3. STRUCTURE AND COMPOSITION

Figs. 26–31 illustrate the structures of typical plasma-sprayed deposits. The coating consists of a heterogeneous mixture of sprayed material, oxide and pores. Shrinkage and degassing after deposition are the principle causes of the porosity but, under suitable conditions, the latter can be kept to as little as 1% of the bulk density of the coating material; on the other hand, bad spraying practice can lead to a condition of gross porosity, as Fig. 32 shows. Oxide inclusions arise either from pick-up during spraying or from excessive surface oxide on the starting powder. As already explained, the former can be controlled by using a shroud or operating within an inert gas chamber, while the latter requires attention to powder quality and particle size.

Table 8. Plasma sprayed materials

METALS AND ALLOYS	CERAMICS	PLASTICS
Aluminium	Alumina	P.T.F.E.
Chromium	Chromic oxide	Ekonol
Copper	Titania	
Iron	Zirconia	
Molybdenum	Chromium carbide/nickel	
Nickel	Tungsten carbide/cobalt	
Silver		
Tungsten		
Zirconium		
Cobalt		
Titanium		
Tantalum		
Aluminium bronze		
Stainless steel		

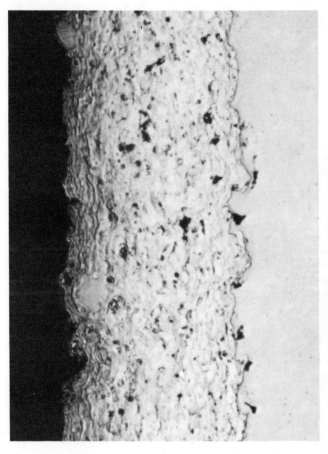

Fig. 26. Plasma sprayed deposit of chromium. (×120.)

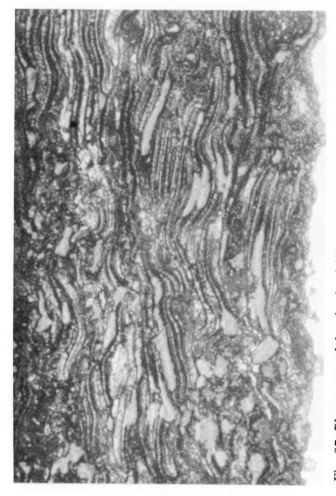

Fig. 27. Plasma sprayed deposit of molybdenum. ($\times 300$.)

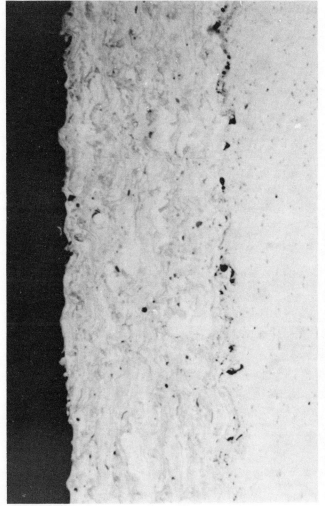

Fig. 28. Plasma sprayed deposit of nickel. (×120.)

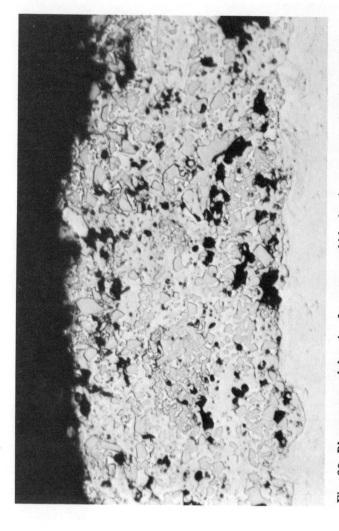

Fig. 29. Plasma sprayed deposit of tungsten carbide showing some porosity. (×240.)

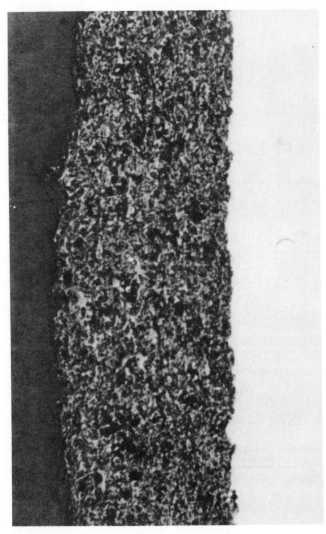

Fig. 30. Plasma sprayed deposit of zirconia (ZrO$_2$). (×120.)

E

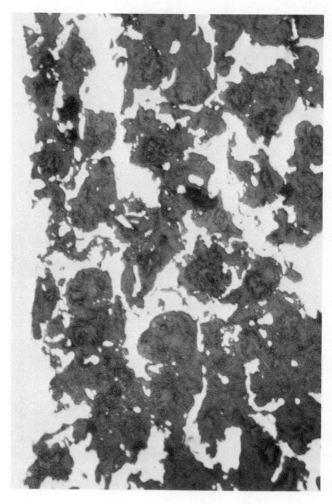

Fig. 31. Plasma sprayed deposit of aluminium/Ekonol. (×250.)

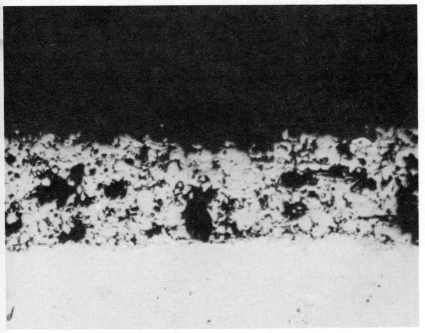

Fig. 32. Plasma sprayed deposit of silicon showing gross porosity. (×120.)

The particles undergo rapid cooling on the substrate and there is insufficient time for significant diffusion. The coating has an undulating appearance and contains fine, columnar grains; equiaxed grains are rarely found. The structural characteristics can be modified by external heating or cooling of the substrate but only to a limited extent. A common structural feature of poor coatings is the presence of unmelted particles (Fig. 33), which indicates incorrect powder quality or spraying conditions. The need for a reasonable temperature interval between the melting and boiling point of the sprayed material has already been mentioned; if this does not exist, excessive vaporisation can occur leading to 'sooty' loosely bonded deposits.

The composition of the coating should correspond to that of the original powders, but it can differ because of decomposition of the material, reaction with the arc gas, oxidation and, possibly, electrode contamination. These can generally be controlled by attention to spraying conditions.

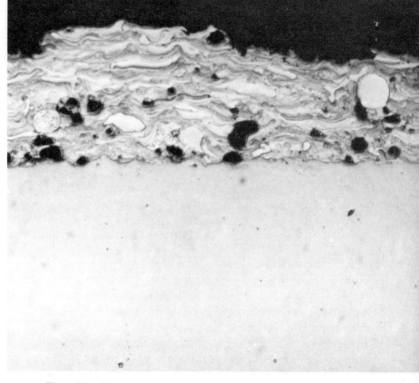

Fig. 33. Plasma sprayed deposit of copper showing unmelted particles. ($\times$ 120.)

It is possible—because of the high cooling rates—to obtain non-equilibrium structures that cannot be produced in conventional material; because of their high free energies, they react rapidly during any subsequent heat treatment, tending towards the equilibrium structure. This can, however, be useful in the preparation of free standing refractory objects since they can subsequently be sintered at temperatures substantially below their conventionally pressed powder counterparts.

6.4. DENSITY AND POROSITY

The precise density of a plasma coating is greatly dependent upon the spraying conditions; Table 9 shows values for a selection of coatings. In general, the porosity is in the range 1–10% (compared with 10–15% for flame coatings); the level can be decreased by subsequent infiltration or heat treatment.

Where it is desirable to introduce a controlled degree of porosity, the plasma equipment is set so that the particles reach the substrate at a slightly lower temperature or velocity than optimum, to prevent them deforming extensively around surface irregularities; this may be achieved either by reducing the dwell time in the gun by introducing the powder into the arc outside the gun nozzle or by increasing the torch-workpiece distance.

6.5. RESIDUAL STRESS

In common with other deposition processes, plasma sprayed coatings may contain residual stresses. These arise from contraction during cooling and they can cause surface cracking or fracture of individual particles. The magnitude of the stresses, therefore, depends upon the thermal expansion coefficient of the coating material; the thermal expansion value may not coincide with the figure for bulk material. For thermal cycling applications, similar thermal expansion coefficients in substrate and coating are desirable. In thin coatings, stresses are rarely sufficient to cause trouble but thick coatings may become detached in service. When difficulty is expected, post heat treatment may be used to relieve the stresses.

6.6. ADHESION

The nature of the bond between the coating and the substrate is not fully understood and is still a matter of dispute. It is probably largely mechanical and this explains the

Table 9. Property data for typical plasma sprayed deposits

MATERIAL	DENSITY, gm/cc	ADHESION, MN/m²	MACRO-HARDNESS, HV10	MICRO-HARDNESS, GN/m²
Alumina	3·97	—	600	15·2
Chromium	7·2	20·7	230	4·9
Chromic oxide	5·2	~30	810	—
Copper	8·92	12·9	85	1·5
Cast iron	—	40·7	—	1·4
Molybdenum	10·2	30·6	230	4·7
Nickel	8·9	24·7	160	1·4
Nickel aluminide	6·0	21·6	160	—
Stainless steel	8·02	42·3	200	2·1
Stellite	8·2	41·6	610	7·9
Tungsten carbide/ 14% cobalt	14·9	> 50	490	25·0 (WC)
Zirconia	4·8	~30	—	5·2

These values should be regarded only as an approximate guide.
Adhesion data obtained with usual epoxy test technique.

importance that workpiece preparation assumes in achieving good adhesion. It has been argued that there is a degree of metallurgical bonding arising from diffusion across the deposit/substrate interface but, in view of the very rapid cooling of the deposited particles, this would appear to be very limited in extent. Microstructural examination of the coating yields a qualitative assessment of the quality of the bond; Fig. 34 shows a coating which clearly possesses poor adhesion.

Bond strengths as high as 400 MN/m² have been reported in the literature but the more usually published values are in the range 25–40 MN/m². Table 9 includes typical adhesion values for a range of coatings. In view of the uncertainties in test procedure, however, published bond strengths should be regarded with some caution.

The bond strength may be increased by various methods.

A greater degree of metallurgical bonding can be achieved by the interposition of a suitable metal (usually molybdenum or tungsten) between the substrate and the (final) deposit. In general, bond strength may be improved by subsequent diffusion either by torch fusing or by post-sintering; however, care must be taken to avoid the formation of brittle intermetallic compounds.

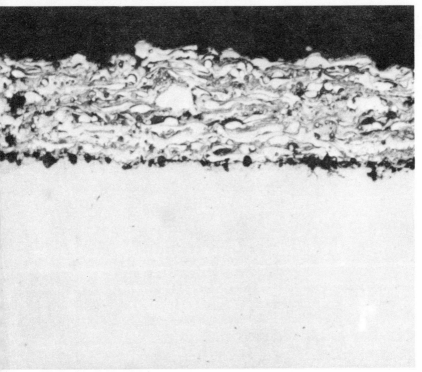

Fig. 34. Plasma sprayed deposit of stainless steel showing poor bonding. ($\times$ 120.)

6.7. HARDNESS

The hardness of the deposit is a commonly quoted parameter and typical values are shown in Table 9. The normal macro-hardness (e.g. as determined in the Brinell test) can be used as a general indication of coating quality for a given material; the harder coating in a group usually indicates the optimum gun/nozzle combination. The presence of porosity lowers the bulk hardness compared with the value in fully solid material and care must therefore be exercised to ensure that a representative hardness value is obtained. The micro-hardness reading, however, can give the true hardness of the individual constituents and is therefore of complementary value to the bulk hardness.

6.8. WEAR PROPERTIES

Since one very important use of plasma coatings is in tribological applications, it is perhaps surprising that very few wear data have been reported. The results of a series of unlubricated wear tests for different coatings against cast iron are shown in Fig. 35. The wear of the cast iron bearer against electroplated chromium was taken as unity.

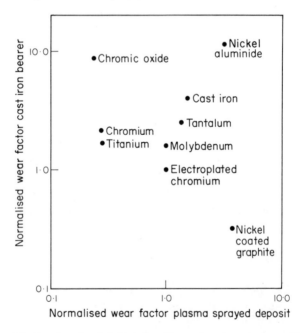

Fig. 35. Results of unlubricated reciprocating wear tests on plasma sprayed coatings.

7. Applications of Plasma Spraying

7.1. ECONOMICS AND AREAS OF APPLICATION

The range of application of plasma spraying is in practice defined as much by economics as by technology. The initial cost of a plasma spraying facility with ancillaries is higher than for a flame spraying facility, although the running costs are not significantly different. However, the high initial cost indicates that plasma spraying should be basically complementary to, rather than competitive with, other forms of thermal spraying. Plasma spraying is selected initially for its high temperature, flexibility, low contamination, low heat input, ability to produce denser coatings, or some more specific feature. Even so, many of the applications described below could be undertaken by other forms of thermal spraying; however, the small cost penalty of plasma is accepted to gain the increased control and superior deposit quality. Table 10 lists some typical applications, and Figs. 4 and 36 illustrate a selection of plasma sprayed components.

7.2. RECLAMATION

One important application of thermal spraying is reclamation, in which a worn or wrongly machined component is built up oversize so that it can then be machined to size. High integrity components are reclaimed in this way by plasma spraying. In the aircraft field especially, the plasma technique is specified for its low heat input and the high quality of the coatings. The aircraft main landing gear axle shown in Fig. 37 is a typical example.

Even the metals of relatively low melting point are often deposited by plasma in this type of work; typical examples

Table 10. Areas of application of plasma spraying

1	Reclamation	Worn components
		Undersize parts
2	Tribological	Porous bearings
		Wear-resistant surfaces
		Self-lubricating materials
3	Chemical barriers	Oxidation resistance
		Corrosion resistance
4	Electrical applications	Resistance heating
		Thick film circuits
5	Miscellaneous applications	Sputtering targets
		Free-standing objects
		Manufacture of special powders, cermets and fibre-reinforced materials
		Plasma sprayed abrasives

Fig. 36. Selection of plasma sprayed automotive components.

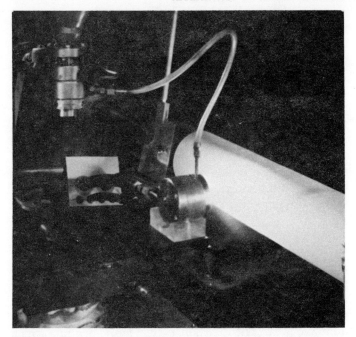

Fig. 37. Plasma spraying of aircraft main landing gear axle.

are the deposition of aluminium on compressor housings and the coating of trunnion balls with aluminium-bronze. The use of tungsten carbide/cobalt is much favoured for the repair of wearing surfaces and this has been so successful that plasma spraying is now being written into original equipment specifications. In the reclamation of highly stressed shafts, plasma sprayed molybdenum is favoured for its high adhesion to the substrate, while chromium is also widely used. Such high quality work will always be a customer of plasma spraying, but it should be regarded as an ancillary rather than a major use of this powerful tool.

7.3. WEAR

The list of tribological uses of plasma sprayed deposits is growing rapidly. In these cases, the main area of interest is the surface, and the substrate requirements are usually efficiently met by ferrous or non-ferrous materials. Materials with satisfactory bearing or wear properties tend to be expensive and may not have optimal substrate characteristics; it is thus doubly undesirable to manufacture such a component in conventional monolithic form. The obvious approach is to apply a surface coating to an efficient substrate. Plasma sprayed materials are particularly useful in this area of application due to their high bond/cohesive strength, the range of materials which can be deposited, and their porosity. The capillarity of the interconnected pore system in the deposits ensures that, even under conditions of conventional lubricant breakdown, a fluid film exists and wear of both members of the materials couple is minimised.

As an example, plasma sprayed coatings have achieved wide utilisation in piston rings for automotive engine applications (Fig. 38). Whilst the pores are an integral part of the successful function of such coatings, the chemistry of the deposit is also important. Chromium deposits in piston rings show some tendency to scuff, an adhesive wear phenomenon, while molybdenum, which has excellent scuff resistance, may become brittle under certain conditions of service. Although both coatings have been widely used in Britain and the USA with success, work is currently proceeding to produce even better deposits. Compatibility of ring and cylinder bore is the major problem since it is found that low wear in one can result in high wear in the other component. However, it is felt that the optimum answer may lie with the quantity and distribution of the porosity which, in plasma spraying, can be consistently controlled to very close limits.

Bearings have been an area of interest in the spraying field because of the ability to deposit non-equilibrium structures. Metal/plastic composites with self lubricating properties have been manufactured by plasma spraying; one example of this approach is a prototype flap trolley plate which has been sprayed to produce a sacrificial self lubricating bearing of aluminium–bronze/Ekonol, a proprietary aromatic polyester with self lubricating properties. Aluminium/PTFE has also been used for this type of application.

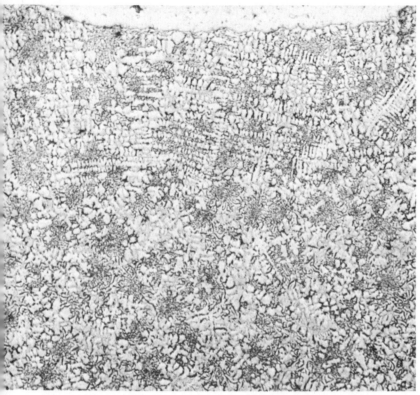

Fig. 38. Section through a piston ring, showing plasma chromium inlay. (×45.)

For pure wear resistance, plasma coatings are pre-eminent since this is a field where the highest melting point ceramics are widely used. Conventional hard facing alloy coatings are being applied to sealing rings manufactured in mild steel to replace parts previously manufactured in one piece from an expensive alloy similar in composition to the hardfacing alloy.

Thin, wear resistant coatings have been applied to the edges of industrial knife blades to improve blade life. In one instance, in the plastics industry, the deposit allowed a softer blade to be used, so that the danger of brittle failure was reduced. The wear resistance of ceramics sprayed by the plasma process is being used to an increasing extent in

73

this area, a typical example being parts of cement process equipment, the surfaces of which are exposed to the highly abrasive hot cement.

7.4. CHEMICAL RESISTANCE

Plasma sprayed materials have only a restricted role in respect of corrosion and oxidation resistance; thermal deposits are not ideal for these applications, since ingress of fluids may not be entirely prevented. However, cathodic protection can be produced in corrosive media by suitable choice of deposit material. Examples of this type of protection include zinc sprayed onto ferrous substrates for aqueous environments and onto aluminium-silicon alloys for protection against aqueous chloride attack; the latter can be a problem in the pistons of certain internal combustion engines. For complete protection against corrosion, sealing of the coating with varnish or epoxy resin must be practised. Deposits so treated are capable of withstanding temperatures approaching 600 K and, in the case of the epoxies, the adhesion is also improved. For chemical attack above 600 K, plasma sprayed deposits provide a useful finite life solution in many cases.

Many corrosion/oxidation applications involve thermal cycling and, in these cases, care must be taken to avoid the deposition of thick coatings which develop large stresses from thermal mismatch and can cause the deposit to detach itself; the use of thin coatings usually avoids this problem.

7.5. ELECTRICAL APPLICATIONS

There is a growing number of electrical applications of plasma sprayed coatings. Much interest surrounds the spraying of barium titanate to close dielectric and ohmic limits, and the spraying of printed circuits is at the experimental stage. The appearance of thermal-shock resistant transparent ceramics has created an interest in wipe-clean cooker tops with the heating elements underneath; the use of plasma sprayed elements gives intimate contact with the hotplate and good heat transfer characteristics. Spraying of an insulator followed by a conventional resistance material in thin strip form is being examined for the manufacture of such hotplates, the insulator being used since the hotplate material becomes electrically conducting at elevated temperature.

Possibly the most exciting field in this area is the deposition of thick films for integrated circuits. Metals for microstrip conductors, metal oxides and cermets for resistors, dielectric materials for capacitors (including barium titanate for use also in piezoelectric transduction), semiconductor material combinations and ferrimagnetic ceramics are all possibilities. Plasma deposition offers the high quality films produced by current thin film techniques with high deposition rates, but successful application depends on the development of fine powder feeding and spraying facilities.

7.6. MISCELLANEOUS APPLICATIONS

A selection of free-standing shapes made by plasma spraying is shown in Fig. 39. Cylinders and cones are manufactured in this way in materials which are difficult or impossible to form otherwise; subsequent sintering is often used to densify the product. An increasing range of special powders is being produced by plasma deposition. The high cooling rates attainable can be used to yield 'splat' cooled powders of lamellar or spheroidal form with ultra-fine structures.

Fibre-reinforced composite materials have been manufactured using plasma spraying. A layer of the matrix material is sprayed on a salt-coated mandrel and strands of wire or reinforcing fibre are wound on the mandrel (see Fig. 40). The composite is built up with successive layers of wire and plasma sprayed deposit, and segments of the material are finally hot pressed to achieve high

Fig. 39. Plasma sprayed free standing objects.

Fig. 40. Stainless steel wire being wound on a mandrel.

cohesion and eliminate porosity. Composites prepared in this way have been directly cast into pistons, to improve fatigue strength in critical regions. A section through an insert is shown in Fig. 41 and a cross section of the hot-pressed insert material is shown in Fig. 42.

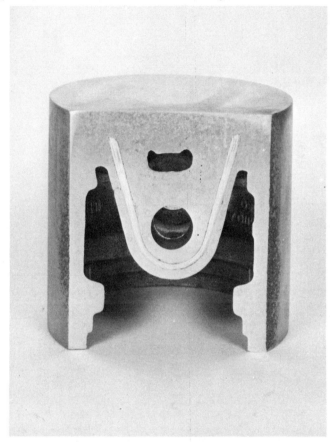

Fig. 41. Automotive piston section, showing a steel wire reinforcing insert.

Yet another promising application concerns plasma-sprayed abrasives; using either a ceramic such as alumina or, for the most sophisticated applications, nickel-coated diamond or boron nitride, cheap but efficient abrasive devices can be manufactured. In the latter field, metal-

F

bonded wheels competitive in price with conventional resin-bonded wheels are envisaged.

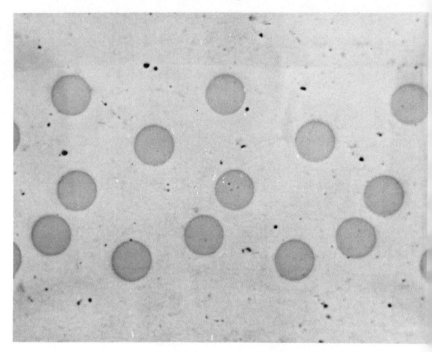

Fig. 42. Structure of an aluminium/steel wire insert. (×75.)

Bibliography

GENERAL

1. W. E. BALLARD, "Metal Spraying" London (Griffin, 4th edition, 1963).
2. J. A. FISHER, "Plasma Spraying", *Met. Rev.*, to be published.
3. H. S. INGRAM and P. SHEPERD, Metco Flame Spraying Handbook, vol. 3: "Plasma Flame Processes" (Metco, 1965).
4. "Welding Handbook", 6th edition, Section 2 (American Welding Soc., 1969). Chapter 29, "Flame Spraying".

BACKGROUND

5. A. MATTING, "Metal Spraying: From Gas to Plasma Jet", Metal Spraying and Plastic Coating Conference (London, Inst. of Welding, 1965).

THEORETICAL

6. F. J. ATKINS, "The Factors Affecting The Plasma Spraying Efficiency of Powders", 2nd Metal Spraying and Plastic Coating Conference (London, Inst. of Welding, 1967).
7. J. D. COBINE, "Gaseous Conductors" (Dover Publ., 1958).
8. H. EDELS, "Properties and Theory of the Electric Arc", *Proc. I.E.E.*, **108A**(37) (1961), 55–69.
9. B. K. SCOTT and J. K. CANNELL, "Arc Plasma Spraying —An Analysis", *Machine Tool Design & Research*, 7(3) (1967), 243–56.
10. J. M. SOMERVILLE, "The Electric Arc" (Methuen, 1959).

EQUIPMENT AND FACILITIES

11. P. G. DUNHAM, "Arc Plasma Generators", Electricity Council Research Centre, Report R100 (1968).
12. A. R. MOSS, "Arc Plasma Technology", Paper 7, Conf. on Recent Developments in Manufacturing Technology (RARDE, Fort Halstead, 1966).
13. "Plasma Spraying—A New Concept in Surface Finishing and Fabrication", *Metallurgia* **81** (March, 1970), pp. 94–6.

TESTING

14. M. A. LEVINSTEIN, A. EISENLOHR and B. E. KRAMER, *Welding J.*, **40**(1), 1961, 85–138.
15. S. J. GRISAFFE, "Simplified Guide to Thermal Spray Coatings", *Machine Design*, **39**(17) (July 20 1967), pp. 174–181.

PROPERTIES

16. M. DONOVAN, "Experiences in the use of Plasma Spraying Techniques". Metal Spraying and Plastic Coating Conference (London, Inst. of Welding, 1965). See also references (4) and (15).

APPLICATIONS

17. D. H. HARRIS and R. J. JANOWIECKI, "Arc-Plasma Deposits May Yield Some Big Microwave Dividends", *Electronics*, **43**(3), (Feb. 2, 1970), pp. 108–15.
18. F. J. HERMANEK, JR., "Coatings Lengthen Jet Engine Life", *Metal Progress* **97**(3) (1970), pp 104–6 (March).
19. R. D. KREMITH, "Plasma Spray Coatings", *Machining*, **117**(3008), (July 1970), pp. 58–61.
20. R. S. KREMITH, J. W. ROSENBERY and P. HOPKINS, "Solid Lubricants Coatings Applied by Plasma Spraying", *Ceramic Bulletin*, **47**(9) (1968), 813–18.

21. M. OKADA and H. MARVO, "Arc Plasma Spraying and its Applications" *British Welding J.* 15, (Aug. 1968), 371–86.

22. H. F. PRASSE, H. E. MCCORMICK and R. D. ANDERSON, "New Developments in Piston Rings for High BMEP Engines", SAE Paper 690753 (1969).

SAFETY PRECAUTIONS

23. C. H. POWELL, L. GOODMAN and M. M. KEY, "Investigative Study of Plasma Torch Hazards", *American Industrial Hygiene Association J.*, **29** (Jul./Aug., 1968), 381–5.

24. A. R. STETSON and G. A. HAVEK, "Plasma Technique for Spraying Toxic and Oxidisable Materials", AIME Symposium Plasma Applications to Metallurgy, 23rd March 1961.

Appendix

MANUFACTURERS OF PLASMA SPRAYING EQUIPMENT*

	MANUFACTURER	ADDRESS
1	Arcos S.A.	Rue des Deux Gares, 58–62, B1070 Bruxelles (Anderlecht), Code 11.262, 00/21–1309, Belgium
2	Associated Engineering Developments Limited	Cawston House, Cawston, Rugby, Warwickshire, England
3	Avco Corporation	75 Third Ave, New York, NY 10017, U.S.A.
4	Metco Inc.	1105 Prospect Ave., Westbury, L.I., N.Y., 11590, U.S.A.
5	Plasmadyne Inc.	3839 S Main Street, D/TR, Santa Ana, Calif. 92702, U.S.A.
6	Thermal Dynamics Corpn.	P.O. Box 10, West Lebanon N.H. 03784, U.S.A.

In addition to the above, a wide range of equipment is manufactured in eastern Europe: U.S.S.R., Czechoslovakia, Poland, D.D.R. Details of these may be obtained from "Plasma Technology" by B. Gross, B. Grycz and K. Miklóssy (Trans by R. C. G. Leckey), Iliffe, London, (1968).

*This list is not intended to be comprehensive. It provides examples of manufacturers of plasma-spraying equipment.

INDEX